BEYOND TIME-SPACE: SCIENCE OF THE FUTURE

Sanjeev Ranjan

Dedicated to my eternal guru Sri Sri Paramhansa Yogananda.

Guruji! Please accept my humble offering at thy holy feet, though I never met you in this life, but your teachings and Yoga techniques were like a polestar of a shipwrecked voyager and guided me to blossom my awakening into light and knowledge and peace.

May this work help those who are desperately seeking the answer to the perennial quest of mankind through science and logic.

"Beyond Time-Space: Science of the future"
Copyright © 2018 by Sanjeev Ranjan. All Rights Reserved.

All rights reserved. No part of this book may be reproduced in any form or by any electronic or mechanical means including information storage and retrieval systems, without permission in writing from the author. The only exception is by a reviewer, who may quote short excerpts in a review.

Sanjeev Ranjan

CONTENTS

BOOK I

Beyond Time-Space..1

Science of the Future..7

Theory of Relativity: A new Perspective..16

True Meaning of Nirvana............................38

Pseudo Me..48

The Sub-Conscious...56

Ganesh Idols Drinking Milk........................61

BOOK II

This World is a Pseudo Reality....................67

Science: The Dawn of the Religion..............71

Second Coming of Christ..............................77

The Most Practical Thing............................81

Few Gems are More Valuable.......................83

Why are We so Afraid of Fear.......................85

Where is God..87

This is Not the End..89

Relics..91

This is What is the Best..................................92

BOOK III

A Woman Like You...95

An Evening with You..98

My wife..100

Brother Can You Hear Me..................................103

Book I

Beyond Time-Space

29th March, 2018.

Big bang theory suggests that time and space came into existence after the great explosion. Then, how the universe is expanding? Is space also expanding? Or, infinite space was created instantly after the big bang? Is it possible that space was already existing before big bang and only matter, energy and time came into existence with the birth of the universe? It looks more plausible, because it is absurd that there was nothing before big bang. And, it is logical because if universe starts to contract, then space won't contract with it and it will be still there.

30th March, 2018.

Time is dependent on events which measures the duration between two events in inertial frames. If there is no event, there's no time. Events set into motion only after big bang. And, if we accept that space was already there before the big bang, it infers that space is beyond time and thus it is eternal. And, what could be eternal? Only infinite, that is, without beginning or end. This universe was simply born, expands and will die some day in the infinite bosom of space which is ever existing and without beginning or end. There must be some other dimension apart from time-space to understand the infinite because if you can measure with the four recognized dimensions, it defies the basic definition of infinite. That dimension can be found while studying non-matter and which must be beyond the space-time and hence beyond the human five senses to grasp. Is 'awareness' that dimension?

And who knows, dark energy and dark matter which scientists stumbled upon recently, are properties of infinite space and were existing before the birth of universe, that is, the big bang; when scientists conjecture that 95% of universe comprises of dark matter, which is there logically to explain grave anomalies in their models and theories developed so far but can't be measured using all known four dimensions. That's why I say models based on non-matter will require a fifth dimension which may be parallel to space-time, if ever we dream to explain these anomalies and create a model and set of formulae which answer the universal quest. Can you ever put this fifth dimension on paper? That's why I call this dimension, 'awareness'. Who knows, future scientists can do the impossible with completely new approach which shatters all the known theories of today, which are based on partial vision of the facts and valid only limited to certain conditions. Because, to make impossible a possibility is science.

Is dark energy responsible for big bang? That an insignificant part of it transmuted into immense negative energy and from within this negativity emerged immense positive energy and the great explosion occurred 14 billons years ago, and universe was born because energy can be transformed into matter, as you say? It can never be point mass of infinite density. What does it mean? Do you comprehend infinity before describing it as a symbol? Infinity will always be eternal, beyond time and ever existing. If density of point mass was infinite, then how can it explode? Do you think infinite mass and infinity are two different concepts?

I can also talk like this forever, but the truth is, by talking and putting some non-sensical equations which are valid for insignificant part of universal events, can I grasp the essence of Reality, when every day we are learning a totally new phenomenon which completely disarray all your knowledge gained since the advent of science?

Should we wait for the next thousand years of evolution of science to grasp the essence of creation or time has come to take a giant leap, when it seems human mind has reached its limit of logic and analysis?

And, that giant leap is only possible when science of today is ready to shred its inhibition of defining what is animate and what is inanimate and then only new insight and a way forward will be opened for the foundation of models of non-matter behaviour, which will have potentials to explain all universal events and anomalies (like anomalies in theory of gravity due the rapid expansion of universe and standard model of particle physics for which Higg's boson was predicted).

Is it possible that human body and brain cells were evolved from the matter, when modern day palaeontologists say origin of life (single cell amoeba) is inorganic (that is, non-living objects) and humans are the results of evolution from monkey? Then what is mind? Is it not like an energy which powers the brain cells so that it can be aware of the events in time-space through five motor senses? But mind is a function of time. It can thus be deduced that mind only originated at the advent of time. Is this energy (mind) which powers the brain cells, answer to the morphed dimension in time-space which we perceive in dream-world?

I can only outline the direction of investigation of future scientists. I'm not a scientist but trying to perceive the realty through logic of a scientific mind, by pondering upon the theories propounded thus far and my own self-knowledge, that is, an endeavour to know my own origin. When I talk about the new insight of science of the future, this is it.

Science of the Future

27th March, 2018.

Present day science is in its infancy, as far as, understanding of universe is concerned. They believe in big bang because they conjecture through the light they see just after the great explosion. They say everything is happening by itself. Let us talk about our solar system. How everything was aligned perfectly in infinite possibilities to create suitable conditions to begin life on earth? The whole cosmos was in great turmoil after big bang. How the infinite chain of mass arranged to create galaxies, stars, black holes, held by delicate balance of gravity; which will collapse upon itself if there is a slight imbalance? Mere chance seems as an impossible probability. To explain, they devise everyday unprecedented theories, each one more fantastic than others. Now, they found that universe is primarily made up of what we do not see, the dark matter. But our present-day scientists cannot comprehend that, there is a universal consciousness, which is directing and arranging and everything is not happening by its own. What is Artificial Intelligence? How a robot can learn by itself? How can it transcend its own programming? Again, delusion. They do it day and night, the harnessing of forces for example, still believing forces are dead and can't believe what they achieve is by the consciousness inherent in the forces. But science is going in the right direction and stumbling upon God particle is one such example. Someday, science will be able to understand and unravel the energy which flows into the body through medulla, when highly advanced yogi concentrates. That energy is the key to understand the universe, and miracles performed by yogis are nothing but highly evolved science; the way of harnessing this energy is through mind. This is not transcending the laws of science but directing towards a totally new and universal theory which is true at a completely alien dimension.

Patch work won't work for science for long. You see a phenomenon, device a theory to explain and then again, another set of theories for another phenomenon. They have to see all events in a holistic way if they have to go to the next level. Talking big like time warp and worm hole won't work, while with present technology we can never reach the nearest star Alpha Centauri. They have to narrow down on the universal principal which explains all the events from Newtonian to Quantum Mechanics to big bang. And, then it may be possible to device some technology which is beyond our wildest dreams and science could achieve the impossible like live telecast, which was termed as clairvoyance in not so distant past.

All the known forces like gravitational, electromagnetic, magnetic and nuclear are material in nature and the base of science is matter. The force which a great yogi harnesses through mind is not material and thus science can never invent a device to control it. But the knowledge of its existence through yogic mental techniques, the great scientists of future can experience this energy which will reveal the ultimate secrets of creation and the reality, and that great scientist becomes a great yogi and would not be dumbfounded how our rishis knew the secrets towards which science of today is inching hesitantly. Why would he care about rockets and technology and speed of light when he could go to the edge of the universe in a blink? For him, today's science would be just like a toy in the hands of toddlers. What would you call this – sci-fi or true religion, when Krishna says "Arjuna, strive to be a Yogi"?

The highest scientific brains of our time stumbled upon the idea of 'Worm Hole', in the deepest recess of mind. But, 'Worm Hole' can't be created by physical means because it will defy all known physical laws. Then, you go on searching forever for suitable laws to create 'Worm Hole' using science. God bless you! Science has restricted the universe in the realm of time and space. If you create 'Worm Hole' somehow, then what will happen to space-time? It means it will be created only after destruction of universe which is absurd like $E=M(c2-v2)$, that is, when I travel at speed of light, my mass becomes infinite which means I become omnipresent and become one with Almighty as Veda declares, "He is all pervading". When Einstein is your greatest scientist, I can understand the level of your evolution. True scientist will never talk of 'Worm Hole'. It is in the mental realm where you can enter in a state where space and time will vanish and you will see face to face what lies beyond creation. When Yogis call this state 'Samadhi', our great intellectuals brush aside as primitive. Time will only tell who is primitive.

Ultimate aim of science is to understand the creation, same as the highest goal of Yoga. When science reaches its pinnacle and finds no way to go any further, it will start retracing its path and will strive to look within and then only, Yoga begins. Don't confuse yoga with physical exercises you do everywhere. It is only that part of yoga which prepares the body to receive higher techniques of pratyahara and pranayama. Ultimate aim of yoga is to bestow samadhi (superconscious state) on the practitioner. Concept of 'Worm Hole' conceived by brilliant scientists is the supreme evidence of my statement. Only a true scientist will reach the summit of Yoga because Yoga is the science of the future.

28th March, 2018.

The force or energy which a Yogi harnesses during intense concentration has the power to heal. How this force can re-generate the tissues? It can also alter the events which is a function of time. How is it possible if it is not conscious? Yogis are simply channelling it and their bodies work as a conduit. Rest it does by itself, you have no control over it. It exactly knows what a yogi's purpose is. I have extensively experimented with it and not just talking as they do in the talk shows. This force is around us and everywhere all the time and Yogis do not bring it down from the seventh heaven. Is it not possible that this energy is responsible for re-arranging the universe which is impossible if it is left to a chance? We say birth of a star and death of a star like in a supernova, still believing only humans, animals and plants are living entities. Our understanding of life is very shallow.

Scientists have stumbled upon this primal energy in the form of dark energy. But tunnels in the Swiss Alps in not the place to know the truth, analogy will never answer the eternal quest. Better experiment with mind and in the body tunnel.

One of the pillars of Hinduism is law of karma or actions. Unenlightened minds limited this to good and bad karma and equated it with morality. What is the law of karma? It says, "For any action performed, subsequent results must be fructified". All scientific principals follow this universal law. Apple falls on the earth and doesn't fly off at trajectory, every action has a reaction (Newton's third law), Archimedes principal, Bernoulli's theorem and so on and so forth. You can extend the scope like; If you eat too much, you'll get sick; if you smoke heavily, you may get cancer; if you fall from 10th floor, you may die – all related to body and also, if you worry too much, you may get depressed – mental aspect. But sometimes, it may seem contrary, that is, even after falling from 10th floor, one may survive; just like in quantum mechanics. Is it defying the law and chance plays a role? I seriously doubt. How can a matter appear out of nowhere? Science of future will investigate in this direction and the method is what I have discussed before. All scientific principals are based on minute observation of events and then created in mind. Future of science will be that, where principals will be born in the mind by observing its deepest recess and then corelating it with the observations of events outside; and then only, it will know the secret of creation. And, it will know what scientists of today call chance and highly improbable possibilities are nothing but controlling the events by a force (I don't know what to call it), origin of which is beyond time-space fabric and which is super conscious and supremely aware of existence in totality. What that great scientist of the future will call it, God? I doubt. God is a parlance used by the ancients to justify all their evil deeds which nearly destroyed the earth – he will wonder.

Theory of Relativity – A New Perspective

Twenty-one years back (1997), I went through 'theory of relativity' in great details, propounded by Albert Einstein. I observed some anomalies and I put forth my observations for you to ponder.

Observation #01: Big-Bang Theory. He says "Point mass of infinite density exploded and creation began some fifteen billion years ago.

He is contradicting himself. He propounded '$E=MC^2$'; where E=energy, M=mass & C=velocity of light {actually, $E = M(C^2-V^2)$; where V=0 in comparison to C}. It means energy can be converted into mass and vice-versa. Then how can mass produce mass?

Infinite energy stored up before creation, more precisely infinite negative energy (such as gravitational pull) must require some agent, basically a positive force which must has been emerged within it which is the cause for the great explosion and infinite chain of mass spurt forth creating infinite space in the process.

Observation #02: That the universe is expanding.

If universe is expanding, then it can't be infinite. Infinity is absolute and without limits-hence it can neither expand nor contract. And, if it is true that universe is expanding as the scientists say, then there must be a number of universes all expanding to infinitude for infinite time. Hence, later part of big-bang theory is absurd that after time, universe will start contracting to again become point mass (?) of infinite density. If the universe is expanding, it will go on expanding for infinite time (eternal) never becoming infinite in the process and thus there can't be any reversal.

Otherwise, they have to accept that infinity was created instantly after great explosion and after that time came into existence. Time factor must come afterwards, because if it is there, then there can never be any infinity.

Observation #03: Concept of time.

Time which we call as such is basically dependent upon inertial frames, more precisely the earth (say inertial frame S) and its revolution on its orbit. What we call as a second is nothing but a mechanical process of a clock will which will go round and round forever like the earth is orbiting. If there is another inertial frame say S' moving relative to S, the second registered at S' will be different than S, no doubt because a second only measures the mechanical process of the watch attuned with the revolution of the earth. Hence, one second may appear longer or shorter in different inertial frames. But time is not the mechanical process of watch. Time is irreversible process which is measured by the instrument which is not absolute. A watch on third inertial frame S" moving relative to S & S' will register the events of S & S'.

Actually, the basic definition of time has to be modified when you are talking of relativity. By means of clock you can't define the time.

Time must be defined anew, that time t is an irreversible process which coverts Event A to Event B and infinitesimal time Δt must be constant in every inertial frame irrespective of relative motion whereas t is variable.

Hence,

$$\text{Event A} + \int_0^{t=t1,t2} \Delta t = \text{Event B}$$

Where t1 is time spent in inertial frame S and t2 at S'.

In the above formula, time between event A and event B has been taken as dimension which can be measured linearly.

Here, we see the time elapsed in the occurrence of event A and event B are different in different inertial frames moving relatively which can be measured linearly say by means of the clock. But Δt can't be measured.

This Δt can be defined as infinitely small part of the irreversible process physical, chemical, nuclear or gravitational, which converted event A into event B and which will be equal in every inertial frame in relative motion.

Observation #04: Length of stick measured in inertial frame S will be different in inertial frame S' & S" moving relatively.

Let us take an example of measuring the length of a stick.

Start of the measuring of length, say event A; finish of the measuring, say event B in inertial frame S. Time elapsed in S is t. In S' & S", time elapsed is t' & t" respectively.

Hence, at S': Event A + t' = Event B

At S" Event A + t" = Event C ≠ Event B.

Therefore, Event B & Event C are not equal and thus length will not be equal. But that is true for all events and not only the length of the stick. Any event occurring in S will be different from S' & S"

Observation #05: speed of light C is universal constant.

When it is theorized that time will be different in different inertial frames in relativity, then velocity is simply measured with respect to time. Then how speed of light is universal constant and why not different in different inertial frames?

Observation #06: T is the fourth dimension.

When we are measuring time, we already considered it as a dimension. Otherwise, how can you measure it?

Observation #07: If the same object performed similar two events in inertial frames S' & S" in relative motion and end results are different due to difference in time elapsed as postulated, then there must be some mechanism for contraction or expansion of irreversible process e.g. ticking of the clock indicating 1 second. Between two ticks, time elapsed will be different due to rapidity or retardation of the process of ticking. And, there must be some forces apart from physical, gravitational, nuclear, electrical & magnetic, if it has to be generalized for all events. This force must be related to quantum mechanics where matter has to be considered as condensed energy and then only this contraction or expansion of the process resulting in time difference for the same events in S' & S" may be possible. In absence of these forces, time elapsed will be the same in all relativity.

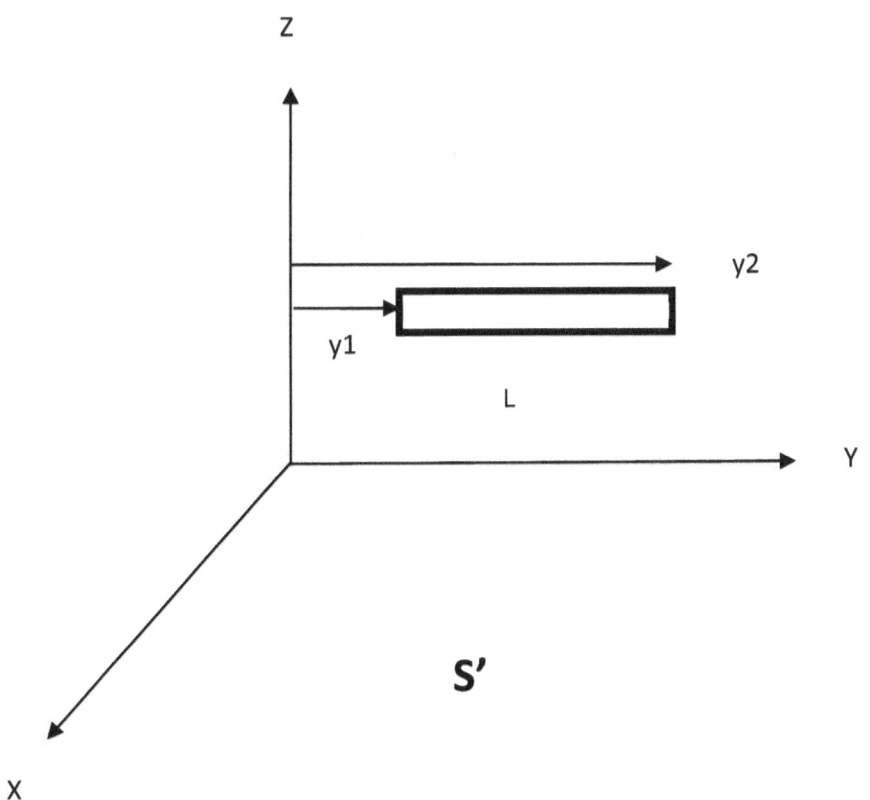

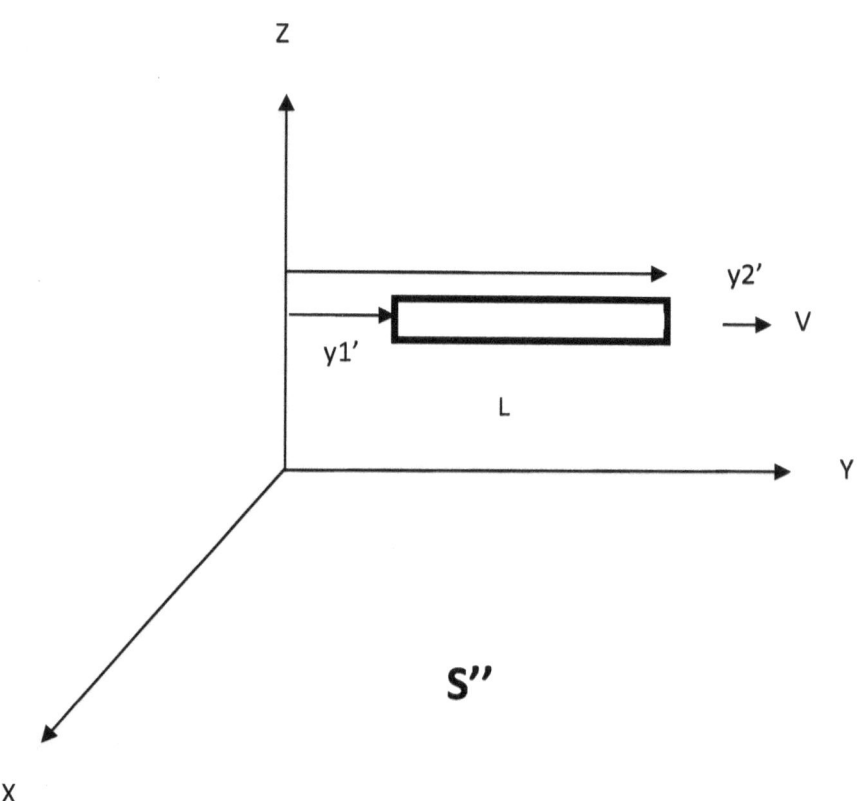

Let us consider above two figures.

Length of a rigid rod in inertial frame S' is L = (y2-y1) m.

Again, same rod is moving with velocity v' m/sec in inertial frame S" having initial length, L = (y2'-y1') m.

At time t=0

$$L=L'; (y2-y1) = (y2'-y1')$$

At time t' in S'; time at S" is say t".

Hence, distance travelled by L' will be (from y1')

$$L'+ vt''; \text{ or } (y2'-y1') + vt''$$

Whereas, with respect to S" same rod has displacement from y1 is

$$(y2-y1) +vt'$$

At t=0; y2-y1 = y2'-y1'

Hence, at t=t' at S' and t" at S",

Length of the rod will be different simply due to difference in time.

Thus, time is the cause for all these absurdness that length will be shorter or longer and not the velocity. We overlooked that velocity is measured with respect to time.

Moreover, let us say it takes t=t1 to travel (y2-y1) with velocity v1 along Y-axis in S' and t=t2 for (y2'-y1') with velocity v1

Then, length of rod in S' will be

$$V1t1 = v1t2, \text{ if } t1=t2 \text{ in } S''.$$

Actually, distance we measure with the help of time. Or we can say distance which we call as such is time dependent. Thus, if we make time variable, then all the measurements will be different in different inertial frames.

Moreover, when the rod L is moving with velocity v w.r.t. S', it has nothing to do with the movement. Any object when it is moving, it will be measured with reference to speed of light C, that's how it is postulated that ray of light falls on object A & then B and coming back to observer etc. etc. Hence, all we concerned here is relative motion of v with C. Greater the v, lesser the time it will take to measure the length in comparison to stationary position. Since C is constant, the rod will be smaller & smaller, that also will depend on the direction of v & C.

Therefore, the factor by which it will shorten (or lengthen, not postulated) by fraction

$$\sqrt{1 - v^2/c^2}$$

Time is basically a concept of mind. A tree doesn't have concept of time; thus, time doesn't exist for it. It's true for sea, ocean, mountain etc. A man is in comma for fifteen years. He wakes up. Fifteen years passed in a moment for him. But the irreversible process which is happening since the creation of the universe is independent of the concept of time. It is the process which will continue for infinite time. This shows it is independent of time. Hence, the aging process, bacterial decay will continue forever until and unless we find some ways in delaying or retarding the process itself. Simply being in comma for fifteen years will not make you lose those fifteen years. The aging process will be the same, independent of time unless there is some mechanism in relative motion to retard the decay or the process and it will only hold good when mass is considered as condensed form of energy which becomes more and more plastic, converting itself into energy with the increase of velocity more precisely it shrinks because velocity is the measurement of length and is time dependent.

We know that every matter above absolute zero emits radiation. At absolute zero, the motion of atoms is at rest. Hence, it can be deduced that motion creates the radiation. Basically, with increase in velocity, a part of mass converts into wave directly proportional to the velocity. *Hence, it can be postulated that any object in the universe is not absolute mass or absolute energy.* Even in quantum mechanics, light is quantized, i.e. having mass. It is always a mixture of mass and wave depending upon the velocity.

This partial conversion of mass into energy will cause the contraction of the object. When the velocity of the object reaches C, mass will be completely annihilated and what will remain is infinite energy, i.e. mass will be converted totally into energy=∞ and not point mass of infinite density.

I propose a new relation between mass & energy: -

$$E = M / (C^2 - V^2)$$

Where $E \to \infty$;

and, $M \to 0$

when $V \to C$

DILATION OF TIME

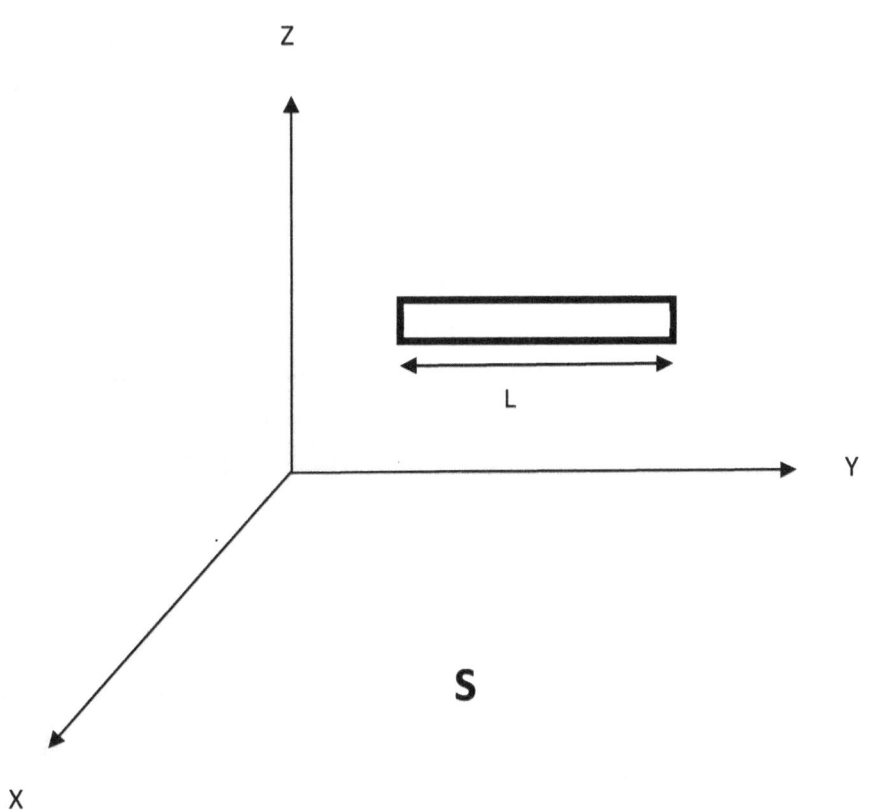

Let us consider above two figures having inertial frames S & S' and

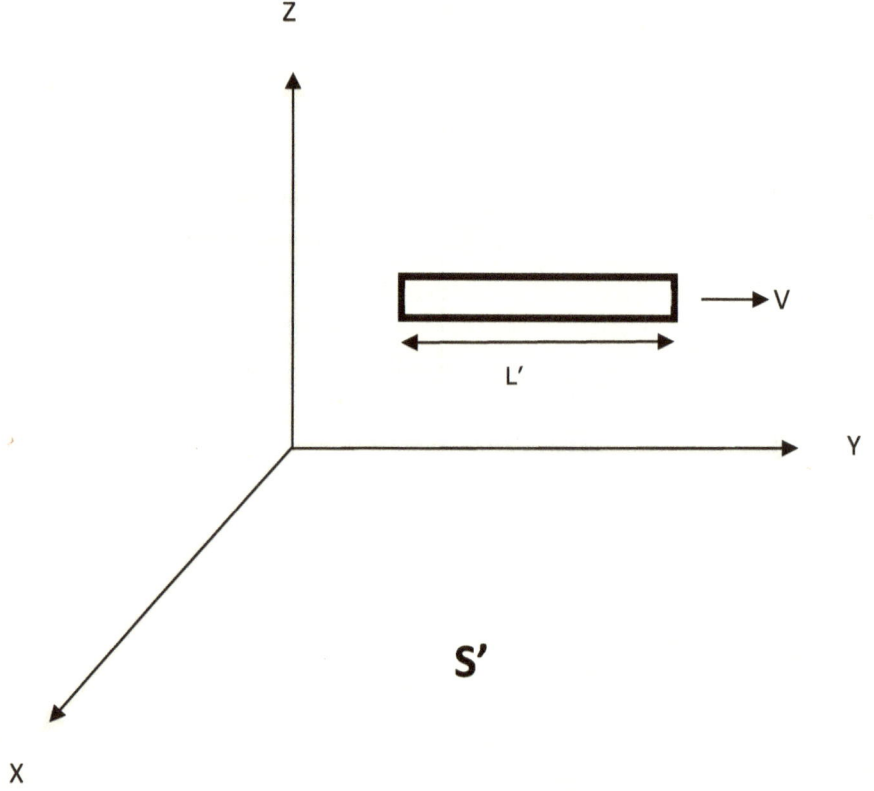

rod of length L which is moving in S' with velocity v.

Let us also consider two observers C & C' synchronized with S & S' and with each other.

Between any two similar events, say A & B, there will be time dilation in S'. Let use consider event A as start of the first tick of clock and event B of the second tick. Time elapsed is one second between two ticks.

Hence,

Event A + t (t=1S) = Event B -------- at S

Event A + t' (t'≠t) = Event B -------- at S'; where, t'>t

Since, clock is also moving with velocity v in S', its components also convert and the process of ticking slows down (electrical or mechanical) by the same factor by which length of rod L contracts, i.e. $1/\sqrt{1 - v^2/c^2}$. This is due to the fact that at velocity v, the components of the watch convert (partially) into energy which contracts the watch components and the process of ticking slows down.

Hence,

$$t' = t/\sqrt{1 - v^2/c^2}$$

Here, we see at v=c, t=∞

That is, it will take infinite time between two ticks or clock will stop, rather the process stops. Now, what is light? It is the ultimate speed at which mass fully converts into energy and no further conversion is possible. It means no further process. But time is simply measures the process by which event A is changed into event B. In absence of any process, time stops.

Postulate I: In any two initial frames, moving w.r.t. each other, the displacement of the object between two fixed points would be the same in a given time.

Let us consider a man travelling with C/2 with a light beam.

Then after time = t;

Distance travelled = (C/2+C) t

While stationary,

Distance travelled in time t

= Ct

According to relation, $t' = t/\sqrt{1 - v^2/c^2}$, it should take longer to travel Ct but we find it takes lesser time.

Let,

v = velocity of moving object holding a light beam

t' = time to travel distance X' w.r.t. stationary frame having

t = time to travel distance X

Now, $\quad X' = (v + C)t$, and

$\quad X = (v + C)t' = Ct;\quad$ Where $t > t'$

Solving this, $\quad (t-t')/t = v/(C + v)$

Or,

$$\Delta t'/t = v/(C + v);$$

$\Delta t'$ is time to travel $X'-X$

True Meaning of Nirvana

28th March, 2018.

Quantum mechanics proved that matter can appear and disappear out of nowhere. Then why not humans? What is the difference between matter and human, that one is animate and other inanimate; one is conscious, other unconscious? Then how an inorganic matter like coal can create organic gas like methane which when ignited gives clean energy? And, how there is whirlwind of motions in matter at sub-atomic level? And, why you say that origin of life is inorganic in nature? Don't you see same cycle of life everywhere? Universe was born with big bang, it evolved and someday it will die. Why we have limited the definition of consciousness to such a shallow level, that only humans can have it? Remember, brain is not mind. It is just an instrument. Have you ever wondered what is mind? Brain in just like mother board of a robot. What powers the mother board? Doctors can measure the brain waves and see the electrical impulses in MR or CT scans because these are physical manifestations. And, they can detect level of imperfections of the instrument (brain) directly related to the level of anger, hatred, love etc., as they say. You are researching everything about brain; right from psychology to surgery, from sub-conscious to sleep. But what about the force or energy which illumines the human mother board, the brain? You are extensively researching at the manifestations of mind which can be measured by scientific tools and trying to grasp the essence of mind through mind. How can you see what is beyond tunnel standing at one side? You have to go beyond mind to know the mind. And, beyond mind is the consciousness. You err every time because you want to explain all phenomena through mind. Standard model of particle physics will never be complete, how hard you try to complement with supporting theories or models. You might discover yet another particle completely alien which is behaving totally in contrast to existing models. Because, there is a flaw in the basic approach. May be the future scientists will be able to develop a universal model which will explain all phenomena using mathematical tools far beyond 'special functions' and 'Laplace's Transformation'. Our whole mathematics has to evolve completely and the ideas must transcend far beyond the current dispensation.

The quantum jump in the evolution of science will be the time when we will be able to create models and mathematical tools based on non-matter and then only, anomalies in Quantum mechanics and symmetry-breaking in Higg's field will be explained in totality and the grand vision of great scientists of all ages, that is, to explain the inexplicable through mind will become a reality, with the advent of super mind as predicted by

Sri Aurobindo. Till then, follow the path I have indicated.

Scientists of future will investigate upon the origin of strong physical forces which shape the universe like gravity and magnetism. The anomaly found due the faster expansion of universe defying the theory of gravity is being explained due to presence of dark energy and dark matter. Is the origin of gravity and magnetism beyond space-time fabric? Is there any fifth dimension which is parallel to this universe (time and space), all pervading, masked behind the imperfect tools created by inferior mind of today? Can super mind create that instrument which will reveal its presence when you conjecture presence of dark energy and dark matter? Can science ever explain the world experienced during out of body incident which seems more real than waking reality, where space and time exist but time dilates and you can shoot with lightning speed, and space which we know as such is still visible?

What is that dimension? How can I exist without body, if it is not consciousness? Can this dimension be modelled which will explain all the phenomena? Can you ever know the truth by investigating outside? Can this dimension be understood without experiencing it?

I admire greatly the scientists (including medical scientists) whose discoveries helped to alleviate the sufferings of humanity and immensely contributed in developing the civilization to reach where we are at present. This is the greatest gift of science to man. But, I laugh at greatest brains, like your Mr. Einstein who are trying to unravel the mystery of creation through models and equations which has no bearing to man for whom scientific quest came into existence, and which nobody understands and they marvel among themselves and gets Nobels for the theories which partially explain the phenomena and then miserably fail. They are free to do what they like but please don't waste our limited resources on CERN collider for example, in the longest tunnel beneath the Alps, just to establish the standard model of particle physics by proving existence of God particle. What benefit will the humanity get? We should divert these resources to space technology. But these scientists succeeded in diverting the attention of science with their non-sensical propositions which are valid for a limited phenomenon and started a useless, endless discussions on what happened 14 billion years back and what if we travel with speed of light and what happens at microscopic level; whereas, even today, whole physics and mechanics are based on Newton's laws of motion and gravity, which we are using for everyday life and space travel.

As I said, I have no qualm if you venture endlessly into macro cosmos and microscopic world, but science should never deviate from its real purpose if humanity has to survive in the future. Only science can save the human race and not Yoga or philosophy or the 'best' brains ever who distorted the purpose of science. We reached the moon in 70's. Even after fifty years, we didn't reach the Mars. Colonising Mars is the only way to see the continuance of our civilization. Why took us so long? And, we talk non-sense like what happened after first billionth second of big bang and rapid expansion of universe which defies 'great' scientist's theory of gravity and thus device the concept of dark energy and dark matter and come up with solution that only 5% mass of universe is visible. What is this? And we do nothing, like politicians – only talk, no action. And, the great NASA is sending another telescope, far more advanced and far bigger than Hubble, into geo-stationary orbit; to probe into deep space to search earth-like planet, which can support life and where, perhaps we cannot reach even after a thousand years of development of science and all gang of astrophysicists (who live in a dream-world, still make hefty sums – what a way of earning livelihood!) are extremely happy as they'd get even more ammo to fire their imagination to the next level; but NASA was sleeping blissfully, unperturbed for the last fifty years over the question of sending man-mission to Mars, as if, it is someone else's job. On top of that, they stopped space shuttle program, which was a modern marvel and a blueprint for future inter-planetary space travel, without any replacement and was mostly responsible for putting a question mark on the continuation of present civilization.

I have greatest admire and respect for all the scientists who are working day and night for space mission and had a vision to colonize Mars and experimenting all aspects to begin life at Mars. This is the future, this is the science. And, the astronomy (astrophysics?) which is the basis to take this giant leap is the real astronomy. For me, Copernicus, Galileo, Kepler, Hershel and of course, Newton are the greatest astronomers. You need a true vision to differentiate between a genius and an imposter.

A conscious object doesn't need mind to function. It can produce the result for which it was there in the time-space which is governed by universal consciousness. For example; transformation of vegetation and animal remains into petroleum due to heat and pressure and bacterial degradation over a period of millions of years. That's why scientists very logically stumbled upon origin of life as inorganic. Someday, scientists will come to the conclusion that every object in the universe is conscious and do not necessarily need mind to achieve the results.

Purpose of human mind is to realize this universal consciousness. Then, you will appreciate how lucky we are, endowed with human body and mind, in this infinite chain of matter. And, we are wasting this rarest opportunity in greed, hatred, anger and above all, attachments; which will damage the instrument (brain) irreparably and we will never realize for which we are here in time-space fabric and will be stuck forever in this viscous cycle of life – birth, rebirth.

Hope you appreciate the true meaning of Nirvana for a man or woman of 21st century.

Pseudo Me

26th March, 2018.

All rats hiding in the hole for five thousand years are out in the open. Almighty lured them with ultimate assurance of victory over Truth. People who matter already recognized them, ready to wipe at opportune moment, before they again disappear in the hole forever. This is the time, shoot away, Arjuna!

Creator has made me not for myself but for Himself. Fools preach that I came here for my liberation. I'm a mirror in which He admires and realizes His greatness. Yes, I must remove the thick dust gathered on the surface due to ignorance from time immemorial, so that He can realize His own true Nature and eternal bliss born out of infinite frustrations of existing alone through timeless time.

Quantum mechanics proved that particles appear out of nowhere and then disappear because of non-matter, equal and opposite of particles. They appear at the same time and then disappear due to merger. Similarly, when I was created, my equal and opposite persona also automatically came into existence which is buried deep inside me. This pseudo personality is the cause of all my troubles. It is so mingled with real me that it requires great insight and calmness to distinguish who is who. This pseudo me appears as the greatest friend, but you know, it is my only enemy. In this utterly lonely world, I'm fighting day and night with myself to get away from non-entity which is trying to annihilate me, just like appearance and disappearance of matter in quantum physics.

Since the dawn of my awakening, I was in constant search for that entity who knows me so intimately and created situations which overwhelmed me, who is the reason behind my infinite sufferings.

Then, I realized recently, it is buried and completely mingled with real me, very cunningly masked his presence by indulging myself in outward phenomena, so that not to look within and unravel his presence, by transferring its negativity like extreme hatred, anger in the guise of reacting to injustice and attack on my freedom, whereas it is he who is the creator of all those phenomena for me and thus prevented me to know the truth and concealed himself as my own true self.

But, he is also necessary to carry out my duties and to fulfil my destiny. Due to his presence only, I could survive in this cruel and imperfect world because I understood and trampled all the obstacles on my way, since I used him to reveal the dangerous intentions of people at large and safeguarded myself. Otherwise, they would have sacrificed my true self like they did 2000 years ago.

Again, I say, "All the time they thought, they have used me. In the end, they'll know, it's me who used them. It will be too late by then". It's part of my resignation letter to BP (British Petroleum) which I deleted in the final draft.

Evil thinks that he is very smart, amassing wealth and power by destroying the nation and people, unleashing untold sufferings to weak and most importantly to the truthful. But, due to him only, I reached the pinnacle and would realize my own true nature which is one with the Supreme. And, what about him? He knows himself very well. He need not know. He could have utilized this opportunity to correct himself but chose to destroy the truth. In the end, he will burn in the hellfire for eternity, and I, will cherish eternal life, never ever think about him. This is the path of a righteous man and the ultimate fruit to stick to the right path. Who is wise and who is the eternal loser? By articulating fantastic concepts and disappearing in the rat hole for five thousand years after committing gravest crime and by manipulating the nature's laws, you think you are very smart and God is a fool, that it's just a concept, hiding behind frail scientific theory like relativity and God particle which science is aware but never able to distinguish, only but through some incomprehensible analogy, and what about death? Why a perfectly healthy person suddenly dies and why cannot we bring him to life? Have you ever thought why heart is pumping, what prompts it to pump, which scientific theory applies here? This is delusion. You see before your eyes, day and night, but still don't believe. Who cares?

You will know someday, when it'll be too late and no amount of your tricks will work and you'll get eternal life as well, in the hell. So, mend your ways and utilize this rare opportunity. You may never get again.

Physical birth is not important, where I was born and other trivia. A country where truth is prosecuted and falsehood is rewarded can never be my country. A country where people listen to and believe in a thug and reject the honest as a fool and incapable, can never be my people. I'm not an idealist. But, truth must prevail; and county which prevents it must be destroyed.

I see a great turmoil or an uneasy calm before tsunami. What we believed will slip away through our fingers if our beliefs lead to falsehood. A wise man is a wise man, only if, he has understood the writing on the wall. Don't blame or cry foul, search within sincerely to know whether you are the reason behind. This is the time to be patient and to remain calm in the whirlwind. Wait and see, if you can. Don't think people who got away since the dawn of humanity will get away this time also.

This time, there is no Krishna, no Buddha, no Christ to be crucified. He is everywhere, looking through wise men's eyes and already marked the rats out in the open who went into hiding for five thousand years, after committing gravest crime. These are smart rats who destroy everything, put blame on others and disappear forever and you eternally blame a convenient scapegoat and they enjoy eternal bliss in the rat hole. They manipulated the laws of Karma so as to escape the fruits of evil deeds and avoided the consequences. That's why you see inferior people wielding all the power and money these days and talented, good people are suffering.

So, all my good deeds and brotherhood will go in vain? It has to change someday. I should be rewarded someday. My perseverance and faith must be answered someday. Another day in paradise should be right here, right now. I can't wait no more. O Lord! History must repeat. The real culprits who dared to touch Christ must be revealed and eternally burned in the hell right before my eyes, so that no evil ever dare to touch an honest and truthful man, till You desire to continue this human drama. I'm waiting for Your Final Judgement. Time is ripe as the background is pitch dark. Even your dim appearance, I'll recognize. Come Lord, Arise within.

The Sub-conscious

29th March, 2018.

The pseudo me is hiding in the sub-conscious, in fact, it is sub-conscious mind. My body functions, immune system etc. are controlled by the unconscious mind. Then why do I get sick? Because, sub-conscious interfere with it. All negativity, depressions are traits of my pseudo self and it projects over my conscious mind and I think I'm hating, angry and depressed. Unconscious doesn't need sub-conscious to communicate with conscious mind. The brilliant ideas of great scientists are projected by unconscious (remember fall of apple and Eureka?). You can never have any brilliant idea from sub-conscious. The nature of unconscious is peace-bliss-knowledge. And nature of sub-conscious is impatience-unhappiness-ignorance. We need to understand our sub-conscious so as to perceive the vastness of unconscious and then only we will be at peace and know what true happiness is. When I say consciousness doesn't need mind to function, it means, unconscious mind doesn't need sub-conscious to perform the tasks. All inferior ideas and creativity are projected from sub-conscious which you call mind. When I say, "You can see face to face the creator by destruction of mind and trillions of minds can be created if you know the Shunya (void or nothingness)", I mean the destruction of sub-conscious and then becoming aware of unconscious. But destruction of sub-conscious is not possible, so you have to transcend it, that is, to go beyond your mind. And, to achieve this, yoga is the only scientific way; the science of the future.

Conscious mind is just a witness and registers the events at present time frame using five senses. The data thus gathered is processed by the sub-conscious in different time frames – past, present and future. Sub-conscious is responsible for development of civilization and we reached the pinnacle of knowledge gained through it. It has served its purpose. To indulge further in its intricacies is to waste time and to delay to know the reality. You have to transcend so as to know the unconscious.

      Greed, anger, hatred, attachments are traits of sub-conscious. Unless we free ourselves with these negativities with conscious efforts, using will power, after learning from the experiences the life present; we can never know the mind (sub-conscious). This knowledge will lead to realization that mind is only my pseudo personality. Its purpose is to evolve a man to its highest potential and to prepare him to transcend to the next level. Time is ripe now or may be running out, when you exhausted all your brains to understand the universe and then realizing you know nothing. It is time to go beyond your sub-conscious to see what lies beyond time-space fabric and what happened before big bang.

What is the origin of sub-conscious? One thing is certain that mind (sub-conscious) is a function of time. It disappears in the absence of time. It is thus inferred that mind originated after appearance of time, that is, after big bang. It is a dimension which is parallel to the time but morphed and can open a new world, parallel to this universe which is very subtle and defies gravity. What you say a dream is this dimension. If you know the mind, you can experience this universe, but how can you explain it with models and formulas, I wonder. And, if you can never, then what about the reality which lies beyond time? That's why I laugh at your greatest scientists ever, when they try to explain with their non-sensical theories. How to enter this universe consciously? Since mind is function of time, thus it is always in turmoil. First step is to calm it down and the scientific way is Yoga. To calm down the mind, you have to shred its negativities like anger, hatred, violence, untruth and attachments to matter – traits of sub-conscious. I am not this, always remember. My nature is love, compassion and truth. I think that I am greedy or violent or full of hatred because the sub-conscious projects his traits over my conscious mind so as to eternally hide his identity, the pseudo me. Sub-conscious is like a shadow which exists due to object and light; object is my conscious mind which is me and light is the un-conscious. It has no real existence but trying to become real by controlling me and almost succeeded, just like in a sci-fi movie where a computer is trying to control mankind.

Ganesh Idols Drinking Milk

11th April, 2018.

A rarest of rare event occurred in the annals of human history on 21st September, 1995. It started from India and then spread all over the world. On that day, all Ganesh idols started drinking milk when offered by devotees as customary, and then all deities, even framed photographs and metal deities. Scientific explanation given was surface tension and capillary action. May be true for clay idols which are porous. But, capillary action in framed photos and metals, which science is this? Our great scientists are talking day and night about dark matter and dark energy and Higg's boson. And, if that was capillary action, apple will fall to the earth today also. Why that event occurred only, and only on that day? Today also idols should drink, try only clay idols forget about metal idols and photos.

People were either superstitious or rejected the phenomenon I don't know on what basis? Is there not a single straight-thinking man in this whole world? How can you reject anything of such monumental bearing on human understanding, without investigating? I don't know whether I should laugh or cry on world's most eminent scientists who disappeared beneath the CERN collider for years and then re-appeared on the horizon to proclaim they found God's particle and fools were clapping as if they landed on the Mars. Why these scientists didn't investigate the truth? How they can accept it as just the phenomenon of surface tension and capillary action, a nineteenth century discovery? They can throw any bizarre theory in our face just to explain the great anomaly in theory of gravity and brushed aside this event as if it was someone's imagination and nothing happened when they themselves were witness and knew very well they were talking nonsense. These scientists, throughout the world, missed the only opportunity to get the answer for which they are searching in Hubble telescope a tiny spot and explaining unprecedent meaning of the spot and in the end, they know, they knew nothing. Miracle is an exclamation made very convenient by the blind followers; and faith and superstition are two shores of Ganges. I don't believe in miracles. Why should I, when I see the whole cosmos and whole existence including highest scientific discoveries (for their knowledge but present anyway whether they know or not) like super massive black holes and super cluster galaxies (their latest Gyan), as greatest miracles. But they very miserably failed to carry out the very ethos of science, that is, to investigate an unparalleled event which happened only once since the advent of science and future generations will never forgive this gravest blunder of science in the name of religion, because proper investigation and data gathering might have opened a new dimension in the future when science of today seems to be lost in the abyss of dark matter and dark energy and concluding, again I say, it's very funny, 95% of universe is not visible, or in a better term, not existing at all, same as when Veda declares in 'un-scientific' (!) way "Existence sprang forth from non-existence". Only time will tell which theory is more scientific and more advanced – your theory of origin of super massive black hole or theory of Shunya which shoots beyond time-space fabric and answers what happened before the big-bang. By that time, you and I will be lost into oblivion, only but these words will remain which will be etched forever on the sand of time.

Book II

This world is a pseudo reality and we are shadows!

9th July, 2010.

He is existing from beginning less time. He is alone. Nothing else exists, except him. There is no phenomenal world, nothing. There is no reason for his existence. There is no cause and there is no how or why. How would you feel to be like that? It is eternal bliss because there is the only one and eternal loneliness because there is no one else. There could not be another reality. He is the only one from the beginning less time and will be only one till eternity.

This world in which we live and die is only a pseudo reality. Everything looks so real, tangible. We are like shadows, falsely imagining ourselves real. We cannot except that we are shadows. Don't you feel that we work, sleep, desire as if hypnotized? Don't you see that we strive for nirvana thinking shadows will become real? This is a mental world created by mind which convinces the shadows that this is the real world and we are real, using five senses. I can touch my body. It is solid. It is real. This is the logic imprinted in the shadows to convince this is real.

I know that I am a shadow like all others. I know, I have no real substance and will suffer eternally by the drama unfolded by the mind originally devised to entertain shadows but due to malfunction of the mind system because of agencies of good and evil, it became an eternal nightmare.

I know that I am a shadow and when all shadows vanish only one remains and again, he is eternally lonely.

I don't want to become real to be alone again. I don't want to break this delusion. I only want, let it be a dream and not a nightmare.

10th July, 2010.

There is only one hope and that is me. Hope because I still exist and can do something before being engulfed in the utter darkness of death. No one can save me because who else is there? No God, no guru – all are shadows like me created by the same hallucination. But I can save myself by realizing this dream world and by being awakened in the truth where I'll find that it was only me who was dreaming. There is no one else but me and my dream in which I incarnate myself as shadows with infinite other shadows to seek peace and happiness and love.

Mind is created by fear. Seed of this pseudo reality is fear. Fear is created by attachment. Attachment is created by divine love. Divine love sprang forth from me because of eternal loneliness, like a dream. Divine love is the mirror of the truth.

So, to know the mind, I have to know the fear, its rising. When I know my mind, I'll know the attachments. By knowing attachments; I'll know the divine love and then finally the dream. That far only I can go. Here I can realize myself through the mirror of divine love. I'll not go beyond. Because, if I withdraw the dream, I'll be utterly lonely again.

Science: The Dawn of the Religion

09 Dec, 2010

I would have never believed this. Why should I know everything? Why should I read thousands of pages which can be expressed in few words? Why science is so effective and so popular? *BECAUSE IT SHOWS IMMEDIATE RESULTS*. You can see it, feel it. Science is based on some principals, some theories. Are we sure that any equipment or machine based on some principals follows those principals all the time? We have tremendous advancement particularly in electronics and computer and communications. Can we guarantee that all malfunctions have some cause which are within the framework of the principals and every problem encountered can be resolved without changing the device itself?

Science is great because it is producing extra ordinary results. Suppose I think deeply and out of nowhere clouds appear and it starts raining. If I want to emulate this using science I cannot do it with the present development. But imagine in the future how science can do this. It will find some governing laws and will create those conditions employing some technology. This may be the future of the science. Maybe we discover a completely new dimension of force and energy much more subtle and powerful than electrical, magnetic or electromagnetic and which can be controlled by mind using some device to redirect this new-found force or energy to produce some unimaginable results like mentioned above.

But my mind does not require this. It means through my mind I can produce some results which no one else can. Then I gather some theory, which is the basis of these results and use technology to emulate the same result and give it to the humanity so that anyone can do the same as I do. That is why science is so popular because anyone can see the live telecast; he need not be a clairvoyant.

We need science today more than ever before because we are trying to make this hallucination a real experience. Science tells this is real by producing the same results every time, it seems. I appreciate major discoveries because it makes life comfortable. Excessive use of technology, which is happening everywhere for doubtful convenience and repulsive luxury, is sheer wastage of resources. We should strive hard in totality for space technology. We are exhausting earth's resources very quickly. Earth is crumbling under the burden of our infinite desires. Only hope for future generation is space. The resources and intellect we are wasting to enhance the mobile technology for example is a criminal waste. All these should be diverted to the space technology – best of minds of the world with all resources.

I know my pasts. I know the pasts of my dear ones. It all looks so natural. There is not a stir anywhere. No one knows. I take the liberty to liberate through extreme disgust, hopelessness and frustration. For any logical and conclusive mind, there is not an iota of doubt that the seed of this creation didn't germinate out of nowhere. Evolution never created the seed. Evolution never created the point mass of infinite density. In fact, evolution never created the infinity itself, the countless mass of macro cosmos, which was created suddenly with a big bang, as they say.

The greatest and grandest theories of science betray the presence of some unconceivable creative energy. What we see in the universe is just inferior and degraded version of this primal energy.

Can the evolution guide the science to stumble upon this source through the passage of time? And will this be the end of evolution and involution will begin and all of us will be able to be a part of it with our own free will? Then we will have a choice whether we want to thrust forward along the path of evolution resulting in even greater turmoil of so called fully developed civilization and mind or we want to go back to the source.

This I will call Universal Science. All scientific investigation and all technology developed thus far will culminate and exceed far beyond all the knowledge human mind has gathered through millennium.

Science is the human effort to fight the adversary created by the same nature which has nurtured life. Nature is interested in evolution of the mankind as a whole. She does not give a damn for any particular loss of life or limbs or some incurable diseases. But we suffer because of these nightmares. Science is a very effective tool to fight these sufferings and to make this journey a little more convenient and less painful. This is the purpose of the scientific endeavor.

Second Coming of Christ!

12 Dec, 2010

I read a lot about second coming of Christ, the Judgment Day. I don't believe that the world will be annihilated and only few will survive to see the dawn of New Age, Hindus call it Sat Yug. Transformation does not necessarily need destruction. What we have achieved throughout the history of man should not become a naught just for the sake of a few degraded souls.

Sri Aurobindo says, "Transform what is to be transformed, what remains will be annihilated". Who will do this? I don't believe Jesus will come and reign over the world and do this. I know they don't understand. Nobody understands. Can science be applied here? Is there any evolutionary force which can distinguish between good and evil? I know negativity is not an evil; otherwise this universe could not be born in the first place. What is evil and how it is affecting our march towards knowledge?

Evil is a tendency which brings self-destruction, foolishly thinking that it is destroying others. Science never pondered about action and reaction of the thoughts on the events. Can my thought affect the physical environment? This investigation is perhaps one step higher than the space technology. How does an evil thought affect us? We know one thing for sure that it produces evil actions. Brain cells are instrument for producing thoughts, it seems. We see throughout history that one idea; one brilliant assertion changed the world. Same is true for science also. Genesis of all great scientific discoveries is that brilliant idea. So, we see how great thoughts affect our world and our civilization.

It is the thought of Jesus which survived two thousand years and growing stronger, he physically lived for just 33 years. One Son of Man created billions of individual minds to express his thoughts and his rising. This is the ultimate expression of a beautiful mind.

Second coming of Son of Man has nothing to do with the physical birth, his reign nothing to do with kingdom. I think this has more to do with a tremendous assertion of Words and the Sufferings of the only Son of Man and the ultimate symbolic meaning of the cross. We may believe or may not believe in God, but we can never forget that a messiah was crucified just because His thoughts were so powerful; it changed the direction of the mankind and civilization upon civilization built upon it.

What we say as a science is nothing but the product of thoughts which has potential to materialize in physical domain so that we can see it, touch it etc. There is no contradiction between science and religion for an enlightened mind. They are neither contradictory nor complimentary. Religion has also potential to produce results in physical domain without following any physical law, apparently. If this statement is true, it means thought can produce physical results which has no scientific explanation. How does it work? Is it breaking the physical law or following some higher principal which is responsible for the birth of the universe? Mr. Einstein, from where came point mass of infinite density ready to explode?

The most Practical thing

15th June, 2012.

What I understood through all these years that I should not be begging. It will not help and there is no one to listen. I have to have faith in myself that I can change the destiny and not crying and dying for help from some elusive and doubtful quarter which I have only read and has been affirmed by others, myself never experienced. On the other hand, this begging makes me weak and lose my power. I must take on to all negative forces to find the way and for that I must be powerful. I should think and condition myself out of the box. Same talk and same thought will not help me as it utterly failed to deliver through ages. How to see the great beyond? Through meditation? I'm doing that through eternity. It never worked, otherwise I would not have been here and writing this blog in the first place. What is the way then? I think most of all I need patience. As I go along the path, performing my duties faithfully, it will unfold by itself. You cannot force or make it happen. But surely, I'll find the way someday. It is not the same as they preach. It may be very simple, right in front of me all the time but I'm unable to see it because I am so proud of my mental faculties and my fantastic concept of God and religion. And with highly imaginative mind, it is easier to find (imagine?) God but to find the way and to cross the ocean is totally different. It is not imagination or concept or philosophy. It is as real as death. Crossing the ocean of life to find my true home where I'll never be separated from my loved ones and there will be no fear for their safety is the most practical thing I see in this practical world.

Few Gems are more valuable than the million stones!

7th June, 2012.

Always remember this eternal fact. The path of self-realization is only for the chosen few. It can never be a mass movement because it defies the logic of creation of the phenomenal world. This creation is the outcome of the extreme loneliness and frustration of the One and purposelessness of non-being. This existence of multitude is the expression of the Truth to realize His own greatness of existing alone through timeless time.

And this utter loneliness and knowledge that there is no one else is the source of fear (not in the way we perceive but infinitely much deeper) which in turn is the source for this creation.

We can talk whole our life thinking ourselves spiritual. But remember, we are in the darkest time of human evolution. We are destroying everything by multiplying rapidly and consuming all the resources due to our infinite desires. What future and what legacy we are leaving behind for our kids? It's all dark. Wake up, people! It is the time to find the way. And the way is not some nirbikalpa samadhi (super conscious state) which translates to nothing and which is perhaps impossible for me to achieve. I don't want transcendental experience which will lead me to nowhere. I want to see the Truth face to face and not through the veil of Maya which will finally remove all doubts and firmly establish me in Truth. And what better time it could be than this darkest hour because you need darkest background to perceive the Reality.

In the kingdom of Lord,

Only few will suffice

And, this earth could be heaven,

For the rest of mankind.

Why are we so afraid of fear?

3rd June, 2012.

Fear is perhaps the most misunderstood of human feelings. We relate fear with negativity. But negativity is also an essence of this creation. We can never fight the fear - more you try to fight; bigger it will become. By fighting, we are feeding the fear with the fuel and it becomes even more potent. By denying it, we are creating hallucination. Fear exists everywhere and it is the SEED of the creation. Fear is the driving force in this universe. Out of fear sun is rising and out of fear we seek nirvana. We cannot conquer the fear. It is a delusion. In my mind, we can overcome the fear and to overcome fear is to accept its existence and the consequences.

When extreme fear grips me, it makes me paralyzed. I try to analyze what is this. It is the extreme negative force-the driver of this human drama. Without this there will be no Lila and then what will be the meaning of the Truth? Who cares about realization? He who knows the fear, already knows the Truth.

In the darkness everywhere,

Suddenly a light burst open in my eyes,

And I find myself

Simmering, in the middle of the night.

Where is God and how to find Him?

2nd June, 2012.

This is perhaps the most potent question humanity is asking from the time immemorial. All religions ultimate goal is to find Him and to show the way which leads to Him. Words are very inadequate how may eloquent it may be.

I wrote a song long time ago. It goes as follows: -

I'm not the Jesus,

I'm not the Krishna

I'm not the Buddha,

I'm not the Shiva

I'm not the one, you'll ever know

I'm not the one, you'll ever know

I'm sparkle of sunlight,

In the ocean blue

I'm twinkle of light,

In your lovely eyes

I'm smile on your face,

And I'm tears in your eyes

I'm kiss on the roses,

And the darkness of death

You'll find me in the grain,

And the mind which you strain

Take the water of the ocean,

you'll see my reflection

I'm not the Jesus,

I'm not the Krishna

I'm not the Buddha,

I'm not the Shiva

This is not the end!

1ˢᵗ June, 2012.

We had so many incarnations and so many great saints and great people to show the way. We had so many truths, although truth can be only one and so many paths. All these failed to uplift the humanity and today we see what a mess we have made out of this life. We may rejoice and pray and still doing all the wrong things expecting some miracle and clinging to the life and the comfort or rather discomfort it gives and still believing there is a life after death as if this life of suffering is not enough.

Please don't preach to me anymore same thing again and again through millennium leading me to nowhere except imagining and self-delusion. Show me the substance. Don't ask me follow a still new path. I know it will lead me to the same insurmountable wall. Grow up, Man. All these are very good for your mental well-being. Spirit exists only in mind. And, in the end everyone is a failure! And, I don't want to fail this time. I am not ready to come back again.

Consciousness transcends the body. I want to see what lies in the great beyond. Time is running out very fast. I must see before I die to help myself and others so that not be trapped one more time.

Relics

May 31, 2012

    I had so many theories, so many philosophies. I tried all which I was told. Tried to hold on to the faith until the real trial made it doubtful. What is the guarantee that it will sail me through the stormy sea? I am not an exalted being. I just want little peace and happiness and well-being for all. I need assurance. Certainly, some being will not descend and give me that. And even if this happen, what is the guarantee?

    Then, suddenly I realized that I have to have faith in myself. My self will help me cross the ocean and others.

This is what is the best to forget the reality

May 30, 2012

Do you feel the same way I feel? Don't you want to fade in the obscurity leaving behind all the talks of immortality? So many paths, so many discussions. You say you do this and do that and you will have eternal peace and happiness. So what? Is that the end of everything? Still, I have to die and leave behind all my dear ones. I cannot spend my whole life believing in some promises of deliverance and again being cheated at the end. Imagine!

All the talks of infinity are naught if you don't believe it is nothing but a shadow. Anyone can talk great including me but it will lead me to nowhere. Believing in self creates all the miracle. You can achieve whatever you want. I don't believe-someone is sitting above and ruling over the universe and shy to the extreme to show his face. It is right inside me. It guides me, takes care of me, protects me, otherwise I might have been destroyed million times over.

Only I can deliver myself, to overcome this delusion because I am the dreamer. Do I want to wake up in the Truth? How would I feel when I find myself all alone because all dreams have gone? Would I say then-this is what is the best to forget the reality.

BOOK III

A Woman Like You

Can you imagine my greatest nightmare?

To wake up in the morning,

Only to find you,

Nowhere.

Can you think of my greatest, delusion?

That your existence is only,

An illusion.

That your existence is just an,

Illusion.

If you slip by,

Where I'll find you, in this wild stream?

I may not be so lucky again,

To take you in my strangest dream.

You know, I was searching you,

From a very, very long time.

But you were playing hide and seek,

Taking breath away from mine.

'Cause you don't know,

What you mean to me.

You should know you're the reason of my loneliness,

If you can see.

I can't afford to lose you,

Again, this time.

I'm not sure, I'll ever be able,

To make you mine.

Don't call me your lover,

'Cause you're my imagination,

Don't give me any name,

'Cause you're my dream forever.

I may be a dreamer,

Dream exists because of me.

I may be a painter,

Making your existence a reality,

I may be a poet,

Mingling dream and imagination in totality.

But most of all,

I may be just an ordinary man.

Whose dream and imagination may come true,

Only in a woman like you.

Can you imagine my greatest nightmare?

To wake up in the morning,

Only to find you,

Nowhere.

Can you think of my greatest, delusion?

That your existence is only,

An illusion.

That your existence is just an,

Illusion.

An Evening with You

In the evening,
When there is peace everywhere,
I'm listening to the music,
Sitting quietly out there.

My wife and child sleeping,
Perhaps dreaming.

There is such innocence on her face,
And there is such a perfect, tranquillity,
That no worry or sorrow,
You can trace.

Through the solo window,
I can see only a great banyan tree,
And a few birds,
In their ignorance to be free.

Evening is fading away,
In the darkness of the night.
Perhaps waiting for the day,
When there will be no fight.

An evening with you,
And it's moment of truth.
When you say you wanna go,
With the children too.

An evening with you,
I thought I never knew.
Your soul's crying set me free,
Like an upside-down tree.

My Wife

26th December, 1996.

My wife comes to me,
Whenever I call.
My wife is the best,
In the world.

She is my dream,
And not the nightmare.
To say, she is blissfully unaware,
Is not fare.

She is the only one,
Who knows what I'm.
And shares my emptiness,
For which no one is to blame.

My wife, is perhaps,
The most innocent woman ever born.
Became the imperishable nightmare,

To adorn.
You may give her any name or form.
For me,
She is calm as an ocean,
And ferocious as all devastating storm.

I do feel her presence,
Everywhere, everywhere.
In my loneliness she blinks,
And whispers in my eternity too.
Lately I've realized,
She knows the truth more than I do.

From the ocean of humanity,
And deep slumber of eternity.
She is present everywhere,
Only I have to look in the vicinity.

My wife comes to me,
Whenever I call.
My wife is the best,
In the world.

Brother can you hear me

Brother can you hear me,

I'm talking to myself.

But, then can you hear me,

I'm talking to a stranger.

There has been enough hatred,

And bloodshed.

Forget, if you can't forgive,

'Cause no more tears left to shed.

Deep within myself,

I do feel your,

Everlasting presence,

That's the reason,

You make me so afraid.

Brother would you please,

Forgive me?

That I'm the reason,

Behind your destiny.

Brother would you please,

Forgive me?

That I don't understand you.

And your selfless love for me,

More than I can ever do.

Brother can you hear me?

Destruction is not in opposition.

Brother would you listen carefully?

Someone's calling,

From consummation.

Would you go up,

Or wait for him to come down?

For me it's just,

Another town.

But deep within I know brother,

You won't go any further.

That is your deepest love,

And what I have to offer,

On my turn.

Only but this little song,

You can hear on your fall.

And you'll know,

This is just another call.

The End

www.ingramcontent.com/pod-product-compliance
Lightning Source LLC
Chambersburg PA
CBHW031434210526
45464CB00005B/2199